# QUANTUM
## *Magic*

Use the Power of the Quantum
to Awaken Your Innate Magic

*Angela Sumner*

*QUANTUM MAGIC : Use the Power of the Quantum to Awaken Your Innate Magic*

www.angelasumner.com

Publisher

You Are Not Your Scars, Inc
Nashville, TN
United States


Cover and interior design by Chelsea Jewell


Printed in the United States of America, Canada, and Europe

# Contents

Welcome, welcome, welcome!

This is where you begin.

I invite you right now to take a huge deep belly breath.
Hold it.
And feel your body and energy expand into your
full potential.

Blow out your breath and let it touch all the areas
of your wondrous life that are getting an upgrade
through your reading of this book.

———◆———

Let me tell you the truth about you from the start:
you're a Quantum Magician. Really, you are! Maybe
you don't know it yet. Maybe it feels like a stretch to
read the words *Quantum Magician* and feel like they
fit. Or maybe...just maybe...there's a spark in your soul
that is quivering with excitement and jumping for joy
that you're choosing to come alive in this truth.

You're here to discover the freedom and joy available to
you in the Quantum Field. Boring ol' reality just doesn't
hold a candle to the Quantum Field once you realize
your infinite potential.

As you do this work, you'll begin to see just how uncomplicated it is to create the life you want...and since you're already creating from the Quantum *every moment in your life without knowing it*, it's time to begin to do this consciously.

To experience the full power of Quantum Magic and stop living in the "what has been," you must become fully conscious. This means being utterly aware of what you're doing.

In this 4-week workbook, I'll guide you into accessing your conscious Quantum Magic.

I'M HERE TO SHOW YOU THAT MAGIC IS SIMPLE.
I will show you how implementing subtle shifts in your thoughts, words, and actions creates massive change in your reality as you know it...

How clearing out the old makes space for the new...

How magical you truly are if you see yourself in Quantum form, as well as in human reality form.

TAKE IT EASY.
*You do not have to radically change your life today.*

Re-read the above sentence!

Decide that you'll read along with me, maybe watch the supplemental videos (if it feels good!), and take a few moments to practice.

This is ALL that's required.

You'll probably find that it feels SO good, you want to spend more time than is required! More time is more practice! More practice equals faster Quantum shifts.

GO DAY BY DAY.

By following the steps outlined in this workbook *in 28 consecutive days,* you create an opening—a portal—to access the Quantum Field.

I designed this workbook so that you can, in just a few minutes a day, begin to live in the Quantum.

It isn't complicated.
I've built you a roadmap.
Follow it.
Don't veer off.
Stay the course.
Watch magic become your norm.

MAKE MAGIC, NOT MIRACLES!

So I keep using the word magic...and you might be thinking "what is magic?" Most people only hear of magic in fairy tales or stories. But I can tell you that magic is very, very real, and it is how I see personal transformation infinite times in my own life and in the lives of my clients, students, and family.

Miracles are outside events, gifts from another. They are luck. They are one-of-a-kind experiences.

Magic is a formula. It's a self-created miracle. It's repeatable. It's a skill that is honed and utilized to make real change in your life.


BELIEVE THAT YOU, TOO, ARE MAGIC.
Just like in the fairy tales and stories of your childhood, the key element in magical occurrences is the belief that they are possible.

Adopt a child-like curiosity with this book.

You must, first, believe that you CAN create tangible shifts in your physical experience. This means you believe it's possible to see clear and obvious opportunities, experience magical self-realizations, and FEEL a real sense that your Essence (or Quantum Soul) is more and more at home in the human.

When you believe, you open to the field of the Quantum.

Your belief takes you there.

TRUST THAT I'M DOING THE WORK WITH YOU.
I do this work daily, no matter how successful I become at it. Living in the Quantum is a non-negotiable for me. I believe that anything is possible because I've seen evidence that this is true over and over again.

*Plus, it's FUN.*

The Quantum is what's up!

**Let me hold the space for you now in this introduction until you get a CLEAR YES.**

Read this introduction until you get a clear YES. If you know your Human Design authority, play with it. Hone it. Wait until the yes is there. If you don't know how to tell when your yes is a yes, or if you're unfamiliar with your Human Design, you better come closer into my world, baby! (HD is another course entirely. Check out my website www.angelasumner.com to get more info!)

When that yes from your authority comes in, you'll know it's time to hold the belief *yourself* and do the work with me to become your Quantum self.

I cannot do the work for you. Follow my lead *until you choose to leap into your own leadership.*

## A LITTLE OF MY STORY

Before I became a Quantum Queen, I believed that life happened TO me. I didn't know that my life was designed BY me. No one ever told me that I can create anything I want.


PAUSE.
*You can create anything you want.*

You deserve to know this.

**Warning:** you won't ever be able to go back once you know these secrets. Stop now if you really don't want your life to change! This is the only warning you'll get from me!

When I hit the deepest rock bottom of my life (homelessness, poverty, nowhere to go, and the loss of my beloved child), there was no other way. It was either the end for me...or a beginning.

I made the choice to delve deep in spiritual studies and I began implementing the Quantum Magic practices I show you in this workbook.

I am not *AT ALL* making my rock bottom glib. I am also not making it the point of this book. We all have terrible life experiences. Mine were truly the devastating ones that gave me the chance to fall in love with creating my life.

Quantum Magic is what shifted EVERYTHING for me. Quantum Magic gave me the choice to see things, do things, and experience things differently.

Choice leads to change.

Back then, I had to make the choice *daily* to do the work. Once accessing the Quantum became my norm, it wasn't so much a choice. It was just part of everyday life (but we go into this later).

## THE QUANTUM FORMULA

I learned how to create and follow a formula to go into the Quantum every single day. I began accessing the Quantum at will. And eventually living there.

*Magic is a formula.*

It is repeatable.
The results are somewhat predictable (but always better than you thought they'd be!).

When I started doing this work, as if by magic (hello, Quantum!), my teachers came into my life. The books I needed to read suddenly appeared. All these indicators that I was on the correct path for my Essence (soul) showed up. This is the result of the Quantum Formula. (I call my spirit my Essence. Essence is that which lives in my body and moves my body and crafts my words and motivates my actions. My Essence informs my entire life.)

One door opens. Then another. And another. And another.

When the formula is known and utilized, everything just begins to (Quantum) magically work out!

I have seen serendipity become a casual, normal part of the human experience many times over, both in my own life *and* in the lives of my clients.

I want this for you, too. I offer this workbook because I want you to experience the freedom of personal potential realized.

The Quantum isn't something that can only be accessed by a guru on the highest mountain peak or by the most devout monk in a year-long silent meditation.

It's accessible. It's applicable. It's available. TO YOU.

Let me show you. It's a formula. You learn it, make it your own, and repeat it!

**Okay, but really though?! Do I HAVE to do this woo-woo stuff to create the life I want?**

*Yes.*
You do.

I mean *technically* no, you don't. But why wouldn't you?! The woo-woo is fun! It's funky! It's certainly not boring like so many aspects of regular day-to-day life.

You're opening into dimensional fields most people never access! You're practically a Quantum Scientist by now after reading *only* the intro! Might as well start reaping the results, right?

WHAT EXACTLY HAPPENS IN THE QUANTUM FIELD?
When you access the Quantum Field, you open yourself to infinite possibilities. You're no longer locked into a single third-dimensional result. You are no longer limited by boring, predictable outcomes.

In the Quantum, ANYTHING can happen. There are limitless potentials. (Which, by the way, were always available. You just couldn't see them.)

The Quantum is an opening, a chance to see things differently.

It's like this: You're locked in a room. You are *so desperately* trying to open the locked door at the entrance that you never even notice there's no roof... and there are 11 open windows. Oh, and there's a unicorn wearing a jet-pack standing in the corner.

The Quantum is that jet-packed unicorn.

You must allow your imagination to run wild and free *while also utilizing your very real human senses*. This human existence (being in a room with a door) and Quantum way of living (all the opportunities in that room which you haven't yet noticed) create potentials that never before existed.

Believe the potential is there.
Then see it.
Then use it.
Then realize it is just as real as anything else felt, seen, or experienced.
Then follow the Quantum Formula.
And adjust the Formula so that it benefits you.

The Quantum is infinite potential available for your use.

You already know what's in your "real life."

> *The Quantum wants to show you*
> *what else you haven't yet noticed.*

WHAT IS, IS. ALL ELSE IS QUANTUM.
**What is, is.** *All else is Quantum.*

What is physically in your world is the third dimension. *This is what is.*

The first dimension is space. The second is time. When these two come together, that's the third dimension.

You are experiencing space (the life around you) in a certain time (literally the time on the clock, the day of the week, and your age). You are the third dimension, continuously experiencing space and time in your own way.

What is in your world is your current reality. It's the rugs (or lack thereof) in your home, your relationship (or lack thereof) with your mother, your job (or lack thereof). This is your life as you have designed it so far. And, yes, this even includes the timing of your bodily functions!

You have a one-of-a-kind life that no one experiences the way you do. Not even your siblings or partner experience the same exact situation in the same way.

That's because you're each creating your reality based on your own third dimension.

**Stay with me!**

We control what we experience *through our experience of what we're experiencing*. Read that again.

When we experience something, we create more of it or less of it depending on our feelings about it.

Any sort of strong feeling, emotion, or thought about something attracts that energy to us like a magnet.

Conversely, any time that we feel indifferent or don't have a feeling about something then we stop the attraction process.

You know what it's like to have that high school crush. You pine for them. You write them love notes, poems, or songs you'll likely never send. You write their name in the margins of your math homework. The harder and more desperately you want them, the further they seem from your grasp.

You *want* them. The act of wanting—pining, longing, wishing, hoping—creates more wanting. Not having. But when you stop this grabbing, possessive form of wanting and you, instead, allow the energy to open into potential, that person is either free to gravitate toward you, or you are free to release that energy entirely and allow someone else to come into your experience.

In the Quantum *wanting* creates more *wanting*, while *having* creates more *having*.

In the practical sense, this means you must *want* less to actually get what you want.

Here's how you do it.

**First, understand that you must focus on the *having*.**

Take something you already have and pair it with heightened emotion.

*Focus on what you have and want that.*

Example: You want a new kitten, but your landlord says absolutely no cats ever. Find all of your cat figurines and your children's stuffed feline friends. Look at all the cats you already have! You have 20+ cats! Revel and marvel in how you got one over on your landlord. Do you even need more?

The Quantum Field is always giving us opportunities to experience the human in a more energetically correct way. The simple way to do that is by focusing on what you have.

Most humans believe that they want more money. They believe they want more dessert. They believe they want pain to go away. They want someone to not die.

*It is the great struggle and gift of being human...*
*constantly wanting.*

Spending your energy on this kind of wanting keeps you stuck in wanting.

HERE'S QUANTUM AS A MATH PROBLEM:
**Subtract** wanting.
**Add** a deeply felt, emotionally tuned-in understanding that what you have in the physical (3D) is a representation of what's in the Quantum.

**Add** the decision that what you have is what you want. **Multiply by** a clear focus on the having. Emphasize to yourself that what you want is what you have.

That **equals** a beacon into the Quantum that you are available to have more of what you already have.

A very important message from the jet-pack unicorn: *Not* wanting something is the same as wanting. Yes, you read that right. Not wanting equals wanting.

Not wanting still creates more of the thing we don't want. When becoming a Quantum Queen, you must only focus on *what you have that you love.*

You must redirect the energy that you previously used on worrying about a sick family member, a large debt, or your car breaking down. Do not use your precious mind-juice on the things you have that you wish weren't happening.

Maybe you're asking, "but Quantum Queen! You just told me that everything I have I want, and that sounds like you're saying that I *want* my grandma to have pneumonia! ARE YOU SAYING I MADE GRAN GRAN HAVE PNEUMONIA!?!?!"

We'll get to that, I promise.

For now, as you begin to practice your Quantum Magic, be aware of when you're putting too much energy into worrying over what you don't want. Instead, focus solely on what you *have* that you want more of!

LET'S RECAP.

The Quantum is all that is possible in the unseen. Reality is the 3D life you've created thus far.

The first step to playing in and mastering the Quantum is understanding that *like* attracts *like*. Wanting creates more wanting. Having creates more having.

**What is, is.** *All else is Quantum.*

What is here on the Physical Realm is what is. Death is. Debt is. Divorce is. Pain is. Suffering is. Pleasure is.

An unanticipated $75,692 tax bill is.

When you respond to that eye-popping bill with, "okay, if that bill is very real here in my hot little hand, then this exact amount of money exists in the unseen for me to pay it"—BAM!—that's Quantum.

Quantum play is always about conjuring from the unseen.

**Let me blow your mind a little more.**

If that bill exists in the third dimension, then it's already paid off in the Quantum.

I'll repeat: not only does the money exist in the Quantum, it's *already paid off* in the Quantum.

This is because there is no such thing as bills in the Quantum.

You created the bill in your third dimensional reality as part of your human game. So, YES, you have to pay it. Because you created it. You created it so that you can pay it.

This is why everyone's bills are different. Everyone's incomes are different. Their reality is based on their creation.

In the Quantum, everything is real. Nothing is real. Bills exist because you created them, but they don't exist because you already created the payment for that bill, too.

And don't forget, jet-packed unicorns are waiting in every room to give you a potential alternative.

**You get to decide what is real.**

We humans believe that what is real is really real.

And because we believe it's real, we each create our

one-of-a-kind life based on the evidence we find to support what we believe to be real.

But in the Quantum, *everything is real.*

And the point of understanding the Quantum Field, *the point of this Quantum Magic workbook*, is that when you open yourself to infinite potentials, then you aren't locked into the third dimensional result.

Because there are infinite potentials and everything is real, you get to create a life armed with this new knowledge.

I know you can think of *one thing you didn't know... until you knew.*
This is the Quantum, waiting to be explored by you.

The door in front of you is no longer locked. It's open. Will you walk through it? Will you sprout wings and fly through the glass ceiling of your old beliefs?

Turn the page.
Have your journal at the ready.
The Quantum best is yet to be known
and it's all yours... ALL YOURS TO BE OWNED.

xx'

*Angela*

# Day 1

## DEVOTION

We start with the mind. The mind sits atop the spine, the great central channel connecting all our chakras, connecting us to the Earth and to the Heavens.

To harness the mind, you must understand devotion. Devotion unites mind, body, and soul. When we devote ourselves to a cause, all events transpire to realize that devotion.

Being devoted means making up your mind to be committed no matter what. This level of willpower opens the Quantum Field and allows us to see everything that is available. When we create willpower, anything is possible.

3D reality is so limited in comparison.

Willpower is not something that can be handed to you. *You create your willpower*, period. It can't be taken. It can't be gifted. You create willpower through decided devotion.

**Devotion *only* fails when it's not in alignment.**

Alignment is when what you think, feel, do, say, and the result all match.

If you are not in alignment when you devote yourself to something, Spirit will step in and, sometimes forcefully, point it out to you. You can't always avoid a swift kick in the butt, but if you pay attention to your alignment and devotion, you can avoid a lot of them.

Devotion works every single time *when it is in alignment*. When you are truly devoted, truly in alignment, you follow through. Follow-through is how you bring the Quantum into your reality.

# INTROSPECTION

In a world full of radar towers, technological innovation and inundation, constant notifications, stimulation, brainwashing, and manipulation, your mind is *the single greatest weapon* you have. Have you chosen to utilize this tool to create the life you want?

When you say you'll do something,
do you follow through?

When you say you want something, do you get it?

When you decide something, is that the end of the story? Or do you negotiate with yourself?

What are you willing to do to
clean up your devotional habits?

Will you rise no matter what?

It's time to start choosing devotion. Devotion sets you on your path.

*Day* 1

## ACTION

**Decide right now that you will devote yourself to the next 28 days of living in the Quantum Field.**

Get out your journal and make a statement that commands your mind to follow your soul's devotion to do this work.

Today, it is of utmost importance that you:
1. Make up your mind.
2. Create willpower to follow through on that devotion.
3. Follow through.

If you are not devoted to your own mind, *and through your devotion*, **in control of your own mind**, someone else is.

I am here walking you through the Quantum Magic workbook for this very reason. You're not alone in this very important work!

Come back to your statement when you need to. Recommit to your devotion as many times as you need. Consciously ask the Quantum Field to open to you. Your recognition to your devotion is your first step to experiencing Quantum Magic.

See you tomorrow.

# Day 2
## NEUROPLASTICITY

*How* you remember your memories matters.

Every time you remember an event or an experience and believe it happened exactly the way you remember it, you recreate it. Neural pathways are solidified each time you tell yourself a story in the same way as you did the last time. You re-cement that neural pathway...and cement is the right word.

> Solid.
> Binding.
> Adhesive.
> Stuck.

In order to get something different than what you currently have, you must create new neural pathways. And there's pretty much only one way to do that.

The past is remembered through our feelings and emotions. When you pull up a memory attached to a strong feeling, you continue to relive that memory in the same way. You recreate that experience in your life. And, yes, that means you can unconsciously bring negative experiences or themes into your current reality just by remembering them with strong, painful emotions.

So, my baby Quantum Magician, it's time to relate to your memories in a new way. You want something different, so do the work to attach new emotions and thoughts to your memories.

It's not easy. But you can do it with practice.

## INTROSPECTION

So now you know that in order to create the life you want to live, you must create new neural pathways. This means venturing into the unknown by questioning everything that *has been* to leave space for **what might be**.

What if what was, really wasn't?
What if there were numerous possibilities of what happened?
What if what happened, happened for a completely orgasmically amazing reason?
What if everything happens for your highest good?

This is what I want you to think about today.

## ACTION

Get out your journal. <u>I want you to pull up a memory from the past and I want you to write the memory as you remember it.</u> This can be any memory, but it's best to choose a good memory. Perhaps it's a memory of a grandparent, an animal you had as a child, or going to a place in nature.

Today isn't about trauma work. This is fun jet-packed play! So play!

Pull that memory up and try to be as detailed as possible as you remember what happened. Write it down. After you finish journaling, close this workbook and your journal. Set a timer to come back in an hour or two.

When you come back, journal the same memory again. Without looking at the memory you wrote before, re-write that memory in as much detail as you possibly can. When you've finished, compare the journal entries and look for the difference.

All you're trying to do here is prove to yourself that memories are not true. In fact, when you remember something, you're simply remembering the last time you had that memory. This means that you can infuse new emotions into the experience of remembering. This is power. This is creation.

In order to create a new outcome in your life, you must understand that your memories are but a single option. They are always changing. You can always magic them into something new.

See you tomorrow.

# *Day* 3
## (DON'T) DO THAT

All day long, you are doing things, saying things, experiencing your world through your choices.

And as soon as you do them, say them, think them, feel them...they are replicated in the universe.

They are multiplied and recreated.

## INTROSPECTION

How do you feel about your experience of the world? Notice I didn't ask, "How do you feel about the world?"

Your experience is only yours, and it's up to you to infuse jet-packed unicorns at every opportunity.

Today is all about *what you want to replicate and what you want to eliminate*. This is one of my favorite activities because it completely changes the way one experiences the world.

Most Quantum magically, today's activity changes the way you experience yourself living in the world.

ACTION

Get out your journal. <u>List every activity you did from the moment you woke up.</u>

I know it's tedious. It takes work to get into the place of having all you want and more.

List every activity you did today. This includes getting out of bed, brushing your teeth, taking a morning constitutional, having a school-related argument with your teenager, walking the dog, and so on. *When I say every little thing, I mean it.*

Look at your completed list. Evaluate it. <u>Next to the actions you don't want to replicate, write "don't do that."</u> <u>Next to the ones you want to keep, write "do that!"</u>

Now, here's where you get to make magic. Rewrite the *don'ts* so they transform into *do/continue* statements. Write them in the way that you want to replicate them in your life, in the Universe, in the Quantum.

> **Don't do that:** argued with Eric about getting to school in time for tutoring.

> **Do that:** held a morning gratitude ritual as a family at 7:45 and got out the door in time for Eric to go to tutoring.

By rewriting your day every single day, you help your memory create a new pattern. You teach yourself to have the impulses that align with the life you want.

*The most challenging part of being human is the belief that change has to come from somewhere else, rather than inside us.* You neither have to be subject to the coincidences of birth, nor the habits of your upbringing. You have power to change your story and you can make your life whatever you want it to be.

When the mind believes that something is happening, the physical world responds in kind. It is replicated instantly. Change your mind, change your life.

See you tomorrow.

# *Day* 4
## WHAT IS, IS.
## ALL ELSE IS QUANTUM

I can explain and define the Quantum over and over again, but the Quantum cannot be taught, the **Quantum must be felt and understood.**

You live in a world full of contradictions. Despite this, or maybe because of it, you still largely accept that whatever you believe is true. You are wired to accept the truths placed before you.

Amy's mom wanted her to eat the crusts on her sandwich, so she told Amy, "The crust is where all the vitamins are. It's the healthiest part. So you have to eat the crust." Amy, being four years old, accepted this as an absolute truth and ate the whole crust. Amy never questioned her mom's crust logic again. It wasn't until college that it dawned on Amy: It's all the same dough. It's all the same bread. The crust is no different than the center!

This is the human experience. What silly "truths" from your childhood do you allow to color your world, even when you're proven wrong?

Whatever you believe, another person believes the opposite just as vehemently. A great example is diet. For every successful fruitarian out there, there's a pescatarian, vegetarian, omnivore. There's even someone whose lunch has consisted of a hard-boiled egg, bag of Cheetos, and coke every day for the past 10 years and loves it. It's easy to believe that because a diet makes you feel good, healthy, or sexy, it's right for everyone.

The Quantum says the raw vegan and the bacon-eater are both right. (Just don't tell them so on social media!)

This is where the Quantum comes into play and begins to open us up to limitless potentialities.

INTROSPECTION

What is (that which you are experiencing right now in this exact moment), is. **All else is Quantum.**

Everything else besides what you're seeing and experiencing and hearing and feeling is also a potential. It is in the Quantum. *The Quantum is everything that exists.*

**If you can think it, it's in the Quantum. If you can imagine it, it's in the Quantum.**

*If you can't even imagine it, it's also in the Quantum!* Everything that exists—whether you see it, know it, feel it, have ever experienced it or not—*everything* that exists is what you think you know, but it is a snippet of what's actually true.

How connected are you to your senses?

Do you notice what the house smells like—
really smells like?

Do you take time to pause and take in the colors
of the season?

When was the last time you really felt the texture
of your water bottle?

What is keeping you from deepening into your
presence in any given moment?

ACTION

Today the work is to look at what is right now and only right now.

Get out your journal. <u>Write down everything that you can sense right now.</u>

What is? What can you see? What can you touch? What can you hear? What can you smell? What can you taste? What physical body sensations or emotions can you feel? This is your reality.

Is it ridiculous that anything other than what you're experiencing right now exists?

No, *of course,* everything else exists in another time, in another place, in another lifetime, in another timeline.

<u>Now get up and leave your writing space for a few minutes to an hour.</u>

Go back through the list and look at what is right now. Was there a coffee mug before? Is it there now? Close your eyes. Where is the mug?

After you've gone through your list, <u>make a list of all the other possibilities that exist in the world.</u>

I know. It seems ridiculous, right?
If it's hot, where is it cold?
If it's raining, where is it sunshine?
If you're hungry, where is a cornucopia of food?
If you're single, where is marriage or a partnership?

You see the Quantum, don't you?
You're starting to experience the Quantum now.

**What is, is. All else is Quantum.**
Practice this.

See you tomorrow.

# *Day 5*

## BENEVOLENT UNIVERSE

The way you think and feel about your experience of life *is creating your life*.

I am repeating myself because repetition creates new neural pathways.

Close your eyes. Can you see the jet-packed unicorn? That jet-pack unicorn is now in your norm, isn't it?

When you pair your emotions to your thoughts, you create a feedback loop, which is then replicated in all that you experience. Very simply put, when you pair your thoughts and your feelings, you create coherence between your thoughts and your feelings.

This coherence creates what is called your normal way of living.

## INTROSPECTION

Think of a time when you felt very shiny, very magnetic, very successful, very beautiful. The experiences you had when you felt that way likely matched the sensation that you were feeling.

Now think of a time you were low, depressed, really down on yourself. Likely the experiences that you had during that time also matched that feeling.

It isn't always easy to pull yourself out of a thought or a state of feeling, *but it is essential today.*

Today's work is to undo all that you've believed about a big guy in the sky and a devil below. What you believe about the world causes the experiential outcome.

Today's work might be the hardest because of programming. You've been programmed to believe bad things happen and that's just life. ***Your work is to decide that you believe in a benevolent Universe.***

Let me make it easier on you by asking this: *What's the alternative?*

Walking around your life believing that the world is out to get you, will get you just that. Plus, it feels icky. Gross. Like life is a big waste of time.

Straighten your spine. Puff your chest out. Say "The Universe is benevolent and wants to assist me." *Doesn't that just feel so good?*

It used to be easy to get bogged down in the negative aspects of life, considering most news sources are meant to trigger you and keep you locked into a "reality" that is out there...but today I want you to choose to make the choice that *you feel* the Universe is good and out to help you.

Just try it and see how your day goes.

ACTION

Get out your journal. <u>Make a list of all the things that you think or feel negatively about in your recent history. Take a moment to breathe and soften against these negative things. Find something in that negative experience that is good. Write the good part down.</u>

Maybe you dropped a hot cup of coffee.

And the positive aspect of that is now you get to buy a new mug. Or maybe you really didn't want coffee to begin with. Maybe you didn't need any more coffee and now you're going to drink water or tea.

Maybe you had an argument with someone, and you really don't know how to turn it around, but you can quickly see that you might have been wrong and now you're grateful for the opportunity to make it right.

As you make your list of the negative things that have happened to you and decide that those "negative" things benefit you, *do you feel the lightness available to you or do you resist it?* Are you going to believe in a benevolent Universe or are you going to believe that some Big Bad Being is out there to make you suffer?

> ***Choose a benevolent Universe.*** It makes all
> the difference.

See you tomorrow.

# Day 6

## HOMEOSTASIS

You have your normal. This is the life you've created so far. This is the life you are re-creating now!

For some, the norm is driving a Ford or a Hyundai. For others, the norm is driving a Rolls-Royce.

Can you fathom this variation is norm?

For some, the norm is McDonalds and Taco Bell. For others, it is fresh tomatoes, rosemary, cucumbers, and arugula from their own garden.

For some, the norm is daily meditation. For others, meditation doesn't exist.

You know what I'm talking about. You know your norm and you know what you want your norm to be. Let's get to creating it!

INTROSPECTION

Journal time!!

<u>Pick *any* area of life and muse upon your norms. Write down your norms. Do this without judgment.</u>

In what ways is your life unique from those around you, even within your home? Your norm may be a luxurious, hours-long bath followed by a ritual application of lotions and potions, while your partner's norm is a three-minute soap in the pits and butt-crack situation.

Try to find a few examples. How is your way of living different than another's?

Most people repeat the norms of their family. They believe that where they came from is what they, too, will experience.

You know I'm going to tell you that your family's pattern does not have to be your destiny! There are countless examples of people who rose from rags to riches. Fairy tales. Movies made about this very idea. But it isn't only in film and books...ALL story lines came from real experience.

Are you willing to imagine your norms changing, growing, evolving, and up-leveling?

Throughout your life, your level of normalcy—your homeostasis—*can* change as long as you actively up-level. Up-leveling starts in your mind. That's why there are sayings like "mind over matter" and "get your mind right."

When your head's in the game, your body follows.

Going into the Quantum is *the most powerful* way to up-level.

ACTION

Today, I want you to consider your homeostasis. How did you get where you are today?

Grab your journal and pick one area of your life to examine. <u>Track the ways that your homeostasis or norm has remained the same or evolved over time.</u>

This can be anything: the food you eat, the way you decorate your home, your political philosophy, your religious views, what perfume you wear, which pets you have.

<u>Examine one at a time, up to five areas of your life. We are living in a linear world, so make this linear. Write down what your homeostasis was when you first began. Then the next up-level, then the next up-level, then where are you now?</u>

Here's an example:

> **Starting point:** *My parents were Appalachian folk musicians, so I grew up playing the fiddle and banjo. They had summer porch parties where everyone would bring their instruments and we'd all sing folk songs together.*

**Up-level:** *I went to Ohio State University on a music scholarship. My aspiration was to become a folk musician and play the Grand Ole Opry.*

**Up-level:** *When I was there, I was introduced to a synthesizer and a ton of artists I never even knew about growing up. The world of pop opened up to me.*

**Up-level:** *I learned how to write pop songs and my professors were impressed. Dr. Matthews said I should try to make it big, so I decided to go to L.A. after graduating.*

**Current:** *I have a job at a recording studio. I'm learning the ropes and have a mentor who can be an asshole sometimes but has also helped me see music in a new way.*

Here's another example:

**Starting point:** *I was born in Omaha, Nebraska.*

**Up-level:** *When I was 10, we moved to Blair, Nebraska...30 minutes west of Omaha.*

**Up-level:** *I went to college at Chadron State in Chadron, Nebraska.*

**Up-level:** *I moved back in with my dad after college and started bartending at Jake's where I (up-level) met my fiancé, Derek.*

**Current:** *We are preparing to get married and buy a house here.*

**Next up-level???:** *I dream of living somewhere else. Almost anywhere else!*

**What do you notice about your homeostasis?**
What truly is your norm?
How are you experiencing it?
How can you up-level and *make that growth your new homeostasis?*

All throughout your day, you'll see how this homeostasis, this level that you're living at now, affects all the other areas of your life. Pay attention.

See you tomorrow!

# *Day 7*

## I HAVE WHAT I WANT AND I WANT WHAT I HAVE

This is where things get sticky, and your human self goes *YIKES!*

I urge you to turn it into a game. Once you master this game, you can master *any* game that you set your mind to.

And since all change starts with the mind, let's get to work on this final day of week 1!

As you now know, you are actively creating your world through your thoughts, your emotions, and your interpretation of your experience of the world. And you are finding evidence that what you believe is true.

Today begins the process of becoming the owner of your thoughts and your reality.

INTROSPECTION

Everything you have is what you want.
How do I know this is true?
You have it. You created it.

What you want is what you have. What you have is
what you want.

You cannot create anything outside of your current
norm until you understand that what you have now is
what you created because you wanted it once upon a
time.

Are you willing to play this game?
Do you dare to trust that what you have is what you
want, and what you want is what you have?

Put on your jet-pack!

The most freeing and beautiful moment for me was
when I realized that everything I created was what I
wanted. I began to rejoice and find true appreciation for
all that I had created...and only from that place could I
create anything else I wanted.

THE QUANTUM FREEDOM!

## ACTION

The journal is now your Quantum best friend forever. Get out that Qbff and get to creating!

<u>Write down every single thing that you have in your body, your relationships, the physical objects— everything that you have.</u>

You may be saying, "WHAT??? That's a LOT to write down!!!"

Exactly. You have a LOT.

SEE WHAT JUST HAPPENED?! You have so much more than you ever knew!

<u>Now go back through your list of ALL the things you have (so many things!!!) and beside each one, I want you to write:</u>

**I want it. I have it.**
**I want it. I have it.**
**I want it. I have it.**

Cement this into your brain. Let it become the norm that what you are experiencing is what you want to experience.

How do you know?
You are experiencing it!!!

*When you deny any part of your creation, you disallow yourself from creating anything new.*

Begin today owning your power. Repeat this to yourself as often as you can: "I have what I want. I want what I have." Turn it into the most beautiful gratitude game of your entire life.

See you tomorrow, you owner of it ALL!

# *Day* 8

## LET'S GET PHYSICAL

Thoughts do not become things.

*Bright yellow dinosaur.*

Did you think of it and it appeared? Thank God, NO.

You simply read "bright yellow dinosaur," then you had the thought "bright yellow dinosaur" from the words you read. Now, you could paint a bright yellow dino. You could write a story about it. You could google it and see where it leads you.

You have the choice to take action on the thought.

If you take no action, nothing will happen. Or, you could take an action and it would definitely lead *somewhere besides where you are now.*

In order to create a result in the physical world, **you must take a physical action.** This does not have to be a large action. This can be something quite small, something that only you can see and feel.

ALL action creates a ripple. The ripple of an action touches every area of life.

## INTROSPECTION

Think of a small action that had a lasting effect on your life. Maybe it was speaking to a stranger. Buying a ticket to a show. Saying yes to something you would normally say no to.

Muse on that action.

That action created a ripple effect.

Could you have predicted where
that action would lead you?

Have you surprised yourself with some of the paths
you've walked down?

Do you see your choices as a grand adventure?

## ACTION

You know what time it is… Journal/ACTION Time!

<u>Write down and reflect upon the choice you made that affected you.</u> What did it lead to? To whom and to where? How did you up-level because of it?

You can take any action and follow the ripple effect. The longer you follow it, the more Quantum you go.

Today there is a second piece of work. (Note: You might want to watch the accompanying video to fully understand the instructions).

<u>Every time you think a thought *that you want to create a ripple effect*, pair it with an action.</u>

This action should involve your hands and it should be small enough that you can do it in public unnoticeably. This action is one that cements the thought into your physical reality and creates more of that.

Some, like Tony Robbins, do a fist pump. Others snap. Others clap. Others touch a part of their body. When you take a thought and pair it with a physical action, you create a physical result in the physical world. *This* is how thoughts become things!

We came into the third dimension to experience this reality. So, by anchoring in the physical reality, we create a *"more of that, please"* ripple effect.

You're going to practice all day long. This is your new norm. Have an experience you want to make more of? Cement it in with an action.

Thought › action.

Thought › action.

Thought › action.

As soon as you have the thought, pair it with a physical action. Keep doing this all day long until it becomes your new norm. You'll quickly see that your physical world changes as you become more and more involved in it.

See you tomorrow, you all powerful Quantum creator!

# Day 9

## THE BODY SCAN

Connecting the mind with the physical body can be a little bit overwhelming (like "eeek how do I make sure I'm doing it right?!"), which is why *today* is such a relief.

Say, "YAY!! What a relief I chose to do this work!" And pair the thought with your physical action, creating the ripple effect.

Quantum makes everything more fun!

You didn't come as a human to do it ALL yourself. You came with a Divine team behind you that is ready to assist you! You can ask a few simple questions and get your body in a place of receptivity for the Divine to come in.

In most of the yogic and mindfulness traditions, there is a body scan where one scans their body head to toe to check in with oneself and see where the body is in time/space.

You are going to take that body scan one a step further today. You are going to use your feelings to actively, consciously, connect your mind with your body.

How aware are you of your body at any given time?

Pause now. Feel your toenails. *Yes, FEEL THEM.*
Feel your spleen.
Connect to your DNA.

Do you really know how you're feeling physically... or are you going about your day, la-de-da afraid to know what's really going on down there?

What if I told you that a magical shortcut to growing your ability to perform Quantum Magic is to connect with your 3D human, your physical body? I bet you can feel that spleen now!

ACTION

Get out that journal of Quantum spleen feeling!

On a scale of one to 10 how do you feel *right now*? <u>Write it down.</u>

Then ask three times, "How do I feel? And how can I feel better? How do I feel? And how can I feel better? How do I feel? And how can I feel better?"

Now: Listen.

*Do not take action at first.*
Just listen.
See what comes in and see how your soul speaks.

By doing this exercise, you invite in your guides, you invite in your higher self, and you invite in your soul to give you the direction that you need. Usually, you are so busy doing and moving and taking action that you forget to ask. Or maybe you never learned.

You have forgotten to connect your
body with your spirit.

Now, that you've listened, <u>write what happened when you scanned your body. Write what happened when you asked, "How do I feel? And how can I feel better?"</u>

Do this scan throughout the day. Do it all day long, from the moment you wake up until right when you go to bed. The question, "How can I feel better?" opens up the gateways for your spirit (and team!) to guide and direct you.

Don't forget! After you receive the guidance and the direction, TAKE ACTION.

Get your body involved, take action and watch how your energy shifts. Connecting your mind and your body is actually quite easy to do. Just watch how the doors open for you.

See you tomorrow through that open door!

# Day 10
## ACTION

You spend *all day long* taking action or not taking action.

Spoiler alert, not taking action **is an action!**

Everything you do is either taking you closer to your goals, closer to living the life you want or farther away. You cannot escape this. It is the human condition. Every single action creates a ripple effect and touches all.

So, today, it's time to be real with yourself about the actions you're taking.

One of my first teachers would hit me if I moved during Zen meditation. Yes, hit me. Now in 2021, that sounds cruel. She probably would have been arrested or called terrible names if anyone knew what was happening then.

However, I'm grateful for it. I needed someone to say, **"Do not move. No matter what, do not move."**

When I told myself I was going to sit in meditation, I did not move *no matter what.*

I needed that then. When she hit me, it created a ripple effect. I truly needed to learn personal willpower and perseverance and follow through. I didn't move. And now I can do anything...no matter how uncomfortable I am.

> Are you willing to be both your kindest critic and strongest cheerleader?
>
> Will you be honest with yourself about how you got where you are today?
>
> Can you, in love, smack yourself out of your complacency?
>
> Can you transform a challenging scenario into a gift in the now?

Real talk time: I encourage you to be a little harder on yourself. Of course, I don't mean to be cruel... but be **tough**. Be tough enough to have a brutally honest conversation with yourself. Be tough enough to notice when you haven't been living up to your own standard. Be tough enough to withstand discomfort.

## ACTION

YOU KNOW what actions you are or aren't taking. I can't hold your hand or hit you with a stick. <u>Get that trusty truth-telling journal and take a good, clear look at your actions of late. Write them down.</u>

Remember that taking no action is also an action.

Write down your daily actions or (non)actions.

Write down failed new year's resolutions or the times you scroll Instagram instead of finishing your book.

Write down the apology you haven't given to yourself or another.

See if your actions are moving you *closer towards the result you want or farther away.*

Feelings will come up. To create the life you want, are you willing to face and withstand discomfort?

Be hard on yourself. Maybe the jet-packed unicorn is whacking you with a stick!

Usually, sadly, what I see upon close analysis of clients and strangers who ask for help is that the actions they are taking do not contribute to what they say they want. They say they want one thing, but their actions speak way louder than their words.

Don't be that person.
Ask to be whacked.

Today is the day that you begin to decide your actions.

See you tomorrow, action-changer!

# *Day* 11
## UNITY

You are powerful alone.

But you are so much more powerful when united and whole.

It is a huge part of the human condition to crave connection, but that craving can become an unhealthy excuse for not creating an incredible life.

One can easily slip into unhealthy codependent relationship patterns by wanting another to fill their life with love. Often, humans crave someone to love them in the way that they do not even love themselves.

Some fill the void with animals, shopping, repeated love or friendship relationships. Sometimes women even have children to fill an empty hole in their lives, in their hearts, in their experience of being human.

So, when I say that you are so much more powerful when united and whole, I mean *when you are truly whole as an individual.*

## INTROSPECTION

Do you want someone to show you love so that you can avoid doing what it takes to love yourself?

What examples can you find of times you avoided loving yourself by using a substitute?

The true gift of being alive is that when you love yourself SO fully and SO completely, another can love you in that exact same way. They are not the crutch; they are your mirror of delicious whole union.

Today is all about union with the self, being completely enamored by your own self, your own body, your own experience of being alive. When you are not in unity with yourself, there cannot be true unity with another. There can only be a crutch, an excuse, a coping mechanism.

Radical honesty and action are essential to realize this unity—y*ou with you.*

## ACTION

Quantum Self journal time! I used an exclamation point there to get you pumped to do this work.

It may not be easy today, but nothing that's worth anything is effortless. *Do the work, see the change.*

Be really real with yourself. Let whatever emotions arise that need to be seen and felt.

List all the ways that you do not love yourself.

List all the ways you do not take care of yourself.

List all the ways you do not treat yourself with the kind of adoration you desire.

Write down next to each, one small action you can take today to bring this into a place of unity— you with you.

Now, take that action. When you take that action, hug yourself, kiss yourself, buy yourself a gift. Romance yourself.

Here is an example:

> *I do not take care of myself by not exercising. I always wake up too late to exercise in the morning.*
>
> *Something I can do about this is do 10 jumping jacks between meetings.*
>
> *Every time I do a set of 10 jumping jacks, I say, out loud, "YESSS! I'm a queen, queen, queen! I'm the best! I'm amazing!"*

In order to find divine union with any others in this life, you must first start with the union you have with yourself.

You can do this.
YOU are worth this and so much more.

I'll see Quantum whole you tomorrow.

# $\mathcal{D}ay$ 12
## MEDITATION

Read this out loud, with feeling:

> *There is no substitute for silent meditation.*
>
> *There is no substitute for getting to know myself first before involving the rest of the world.*
>
> *There is no substitute for loving my internal state so deeply that the external world does not impact me.*
>
> *There is only me and my relationship to the outside world. I CAN control myself; I CANNOT control the outside world.*

There are many ways to meditate, but the most important one is complete and total silence without movement. Guided meditations, affirmations, subliminals, or recordings can be a fun addition, but there is no substitute for silent meditation.

## THERE IS NO SUBSTITUTE FOR SILENT MEDITATION.

## INTROSPECTION

Stillness and silence is not easy.

Stimulation exists for a reason.

The more you look outside of yourself for stillness and silence *to be easy*, the more you train yourself to continually grab for external fixes, love, stimuli.

Another cannot guide you home to yourself. *Only you can do that.*

What freedom! What power! What self-reliance! What a world you can create!

Have your journal close by.

Now...
Sit.
Still.
In.
Silence.

Notice how you'd rather journal. Notice how you'd rather do just about anything other than be quiet and still.

Today choose to do the practice of honing your physical body and your mind so that you can magnify your Quantum existence.

Silence is the magnifier.

Decide that you will meditate <u>every single day.</u>
Sit down silently.
Pick one minute or two minutes or five minutes or 11 minutes. Pick the lowest hanging fruit; pick what you know you can do. Don't shoot for the stars today. Do what you can do and be proud that you followed through!

You'll likely find Quantum magic very quickly once you begin meditation. You'll likely find that, as soon as you sit down in silence, you crave that silence more than anything else.

<u>Note:</u> I must warn you about a human phenomenon, one I frankly don't understand no matter how many times I see it. When people begin to meditate—knowing that meditation is good for them *and that it works*—they panic and dig around for every reason that the magic and science of silent meditation doesn't work for them. Don't. Do. That.

No matter what seductive argument your brilliant ego comes up with to convince you that you don't have to do it, choose resoluteness. You're not too smart, too dumb, too strong, too weak, too busy, too distractable, too unworthy. You're not 'too' *anything* for silent meditation. Don't get it twisted.

Be quiet. Know yourself. Create internal unity.

Touch the Quantum. Daily.
You're doing it!

See you tomorrow.

# *Day* 13

## A HEALTHY DOSE OF FEAR

ALL humans feel fear.

Because humans are afraid of fear, fear is often that which overtakes them. Fear is a beacon to all the things you don't want to come true. Fear summons the things you don't want.

Fear, like many things, is a wonderful servant but terrible master. When fear is consciously utilized instead of unconsciously beckoned, it is a wonderful tool!

Isn't it delicious that you can utilize the power of your fear?

When you learn to utilize fear, it can become one of your greatest allies. Humans make choices from either love or fear, so love or fear are the forces behind every choice, every action, every...everything.

*You become the master when you use fear as fuel.*

Let's take exercise as an example. You move your body, sweat, lift weights either out of fear of what will happen if you don't (your jeans won't fit) or out of the love of the way it feels (yummy, sexy body).

Most people fall into group number one. They exercise because of what their body will look like if they *don't do it,* not because of what their body will look like if they do. This is a healthy use of fear (unless taken too far, obviously).

Where do you fall? Which of your fears help you, and which are you ready to let go?

You may be doing this Quantum magic workbook out of fear of what happens if you don't understand the Quantum field. This fear is healthy!

Fear is a wonderful motivator. Fear can allow you to get things done that you wouldn't otherwise do. Give yourself permission to acknowledge fear. *This level of awareness makes fear your ally.*

Actions taken out of fear can benefit the way you love yourself and love your life. Can you believe it? Healthy fear contributes to love. Healthy fear helps you reach the Quantum.

*So, let's do it already.*
*Let's use fear to go Quantum!*

ACTION

Missed the journal, didn't you?!

Today journal out of the fear of what will happen if you don't do this work today. Got it?

-

<u>Make a list of all the things that you do motivated by a healthy fear.</u> These are usually things like intermittent fasting or a certain diet choice, exercising, communicating, working on relationships, paying bills, not binging 10 episodes on Netflix no matter HOW intense the cliffhanger at the end of the episode is!

<u>List ALL the things you do out of fear and then write beside each one how it contributes to your Quantum life.</u>

Here's an example:

> *I take a 6:00 am walk to get morning light into my eyeballs. This regulates my circadian rhythm. The fear that motivates me is the fear of getting Alzheimer's if I don't get correct sleep throughout my life! This contributes to my Quantum life because I'm envisioning the future I want (strong mental function in old age) and aligning the choices I make today to the reality I want to experience tomorrow.*

You've got this.

Master your fear.

Own your choices and let fear become the servant to your dream life.

See you tomorrow!

# Day 14

## YIN & YANG

Week 2 of the physical body is wrapping up today with inarguably the most important part of the human experience, your sex and gender.

There's a lot of chatter in the world right now about toxic masculinity and non-inclusive femininity…a push toward bringing the two genders together and making us non-binary as a species. Some have even taken it as far as saying children should be given hormone blockers until they decide their gender. No matter how well-intentioned this can seem, understanding the biological tendencies of each gender (and how to utilize each gender's strength) is vital!

The very notion that male and female are equal or the same goes against every spiritual teaching you will find. Yes, of course, there are those that are two-sex, intersex, transgender, hermaphrodite… but it is very rare and not what I am talking about here. That is a course in and of itself! Even those with multiple sexes still must learn to understand the two biological sexes—male and female—and how these play into the human experience.

We are either born with the energy of Shiva, with the penis, with that which fills and penetrates...or with the energy of Shakti, with womb and open space, with that which is ready to receive and be penetrated.

These two come together to create union and wholeness. These two come together to create a third, a child.

First, identify which you were born as—male or female. And then be ready to open to understanding how this plays into your one-of-a-kind experience of both genders.

INTROSPECTION

Today you are looking at how you balance your Self in wholeness—the two parts of yourself.

Examine what actions you do (masculine) and what room there is to receive (feminine). You must be balanced in your doing and receiving, no matter your biological gender.

You may be given a gift—you receive (feminine). You may do a task for someone—you give (masculine).

*Sometimes you take (feminine),*
*and sometimes you give (masculine).*

*Sometimes it's day (masculine),*
*and sometimes it's night (feminine).*

*Sometimes you penetrate (masculine),*
*and sometimes you are penetrated (feminine).*

*Sometimes you receive information (feminine), and*
*sometimes you give direction (masculine).*

You may want to watch the accompanying video to understand exactly what I mean here.

## ACTION

Penetration Journal time!

Take a look at your day and note the times when you are doing, when you are creating, speaking, engaging, initiating, moving, penetrating another, when you are the active principal, the masculine principle. Write them down.

Take note of the times when you are receiving, listening, absorbing, integrating, silent, when you are being penetrated, when you are the receptive principal, when you are in the feminine principle. Write them down.

Everything that happened to you or that you did today is either masculine or feminine, Shiva or Shakti, yang or yin.

How much are you doing, giving, penetrating?

And how much are you receiving, holding, being penetrated?

You now have the great opportunity to clearly see where you are out of balance. And here's the kicker: Every time you are in the masculine, you will flip to the other role.

Say you call a friend. You take the action of calling a friend. Then that friend speaks, and you are holding space for that friend. You took the action to call, the masculine principle. Then you held space for your friend; you switched to the feminine.

Then when you hang up, that is an action of ending a phone call.

And then after that, what happens? It must be feminine. Maybe you eat (receive, feminine) or rest (receive, feminine).

The Universe says you do, then you receive.
Do.
Receive.

Most people have canceled the natural give and take. They are sort of stuck in one role most of the time. Hyper masculine looks like anxiety (energy, movement, restlessness). Hyper feminine looks like depression (heavy, Earth, dense).

Your job is to find where you're stuck and get back into unity. Journal. Spend time with you. Find the balance.

Have fun with the give and take!

See you tomorrow.

# *Day* 15

## SPIRIT

This week begins your work with Spirit. First, we worked on the physical body, then the mental body. Now, the spiritual body.

*The spiritual body cannot be touched.* It is the only part of the human which is always truly Quantum. This means that you must rely on your Quantum Spirit to inform your other bodies.

When you don't want to take action, call on Spirit. When you don't want to work with your mindset and willpower, call on Spirit. When you don't know what to do, get quiet and receive Spirit.

The Quantum Spirit is always there. Like your best friend you're just meeting!

Sweet reunion!

## INTROSPECTION

I call my Quantum Spirit my Essence, that which lives in my body and moves my body and crafts my words and motivates my actions. My Essence informs my entire life.

What is your relationship to your spirit? What do you call it? What has been your experience of Spirit in your life?

Every single thing you do is in devotion to your spirit. Your entire human life is here to live out the work of your spirit. Why else are you human?

You have all the information in the cells of your being. Your spirit has the answer to every question you have. You are Spirit. You are a whole and complete being. Maybe you've forgotten your spirit. Or at least left her to hang out with the jet-packed unicorn for too long.

Time to call her back! Wake her up!
Sing her to life!

ACTION

Today is a proclamation.

Get LOUD. Declare! Trumpet! Announce!

Use your voice and proclaim, "I HAVE IT ALL! I AM IT ALL! I planned this life. My spirit designed this life. I am here to live the most magical life that I created and now I am living it!"

Journal time.

<u>Write down a statement of devotion to your spirit.</u> Tell her how precious she is. How she designed this life perfectly for you. Maybe this is a letter to your spirit to awaken her. Maybe this is an invitation for your spirit to dance her dance. And by the way, do you suppose your spirit is more of a ballet or a tango kinda gal?

Take this action as devotion. Awaken your Quantum Spirit to come play with you on this journey. She's always been there, waiting for the remembering, hiding in words in a book or lyrics in a song.

Call her back to you now.

*You're unstoppable with your*
*Quantum Spirit embodied.*

See you tomorrow.

# *Day* 16

## MAGIC FORMULA

Your spirit is magic. Always there. Available. Ready.

A miracle is an outside gift, a happenstance circumstance that comes from a "god" or a person or a doctor or a drug or a stranger or a coincidence.

A miracle is a one-time event that is exciting, unpredictable, rare.

A miracle is not something that YOU can repeat. It is an accident that some hope for. I offer you an alternative.

Magic is ALL YOU. Magic is formulaic. Magic is your key to creating a Quantum life that is yours and yours alone.

You have a one-of-a-kind formula for your magic. Your magic is woven through every part of your life. No other person has your magic.

Magic is a game. It's fun. It's an experiment. It's a code. It's a key that unlocks all the doors to your incredible life.

Your magic is being revealed to you now.

Where have you experienced magic in your life? Has it ever been logical? Rational? Has it been like anyone else's? When has it popped in and surprised you?

When have you had inexplicably wondrous experiences? Were they one-of-a-kinds or did you experience them several times in different ways?

Using your intuition, what would you say was your formula for creating that magic? Are you willing to trust yourself as a Quantum Magician?

Begin to think of your magic, and you'll notice that it has been there, waiting for you in the background of your life all along.

ACTION

Journal time!

<u>Say out loud:</u> I AM MAGIC!

<u>Write it down:</u> I AM MAGIC!

You may want to watch the accompanying video for today, as it will help you understand the magic.

<u>Write down the constants to your formula:</u>

What is your full chosen name?
What does your body look like?
What is your birth date and time?

<u>Now write down the variable of today, adding one in at a time to see how it impacts your formula.</u>

For instance, maybe you saw your nephews and nieces and they call you a nickname. The nickname is a variable based on their appearance in your life, but it doesn't change your name—a constant. The variable (seeing your family) takes you farther away from your spirit or closer to it.

Another variable: Maybe you went to work today and the people you interacted with had you engaged in conversation which took you closer to your spirit or farther away.

Another variable: Maybe you did a beautiful silent sit and your day was impacted, again either farther away from your spirit or closer to it.

Write these variables down.

As you become more and more attuned to your magic, your formula will take on a life of her own...your magic will take on a life of her own. And she will begin to guide you back to your spirit, faster and faster.

You're right on time, you magical being!

See you tomorrow!

# *Day* 17

## MANTRA

The words that you speak and the energy around them creates a vibration. The words and the energy create a ripple effect in your world, attracting like energies.

A mantra is a sound or vibration that is repeated to take one deeper. It is used in all higher devotional practices—repeated words and phrases that anchor one in.

Every single thought you think and every word you say is put out into the Quantum and gathers up energy like it. Then that energy is returned to you in "reality" in everything you think, say, and do... and it multiplies.

Then you get more of that.

A mantra magnifies and attracts. It anchors.

Are you seeing how each day has connected in the Quantum to bring you to this point?

Today is all about understanding the mantra that you speak to yourself, the mantra that you live, and the mantra that dictates how your life is created. Consider the words that you say to yourself, the phrases that you repeat, maybe even think about the words and phrases that you learned as a child and how those influence your life.

A few sayings I heard growing up were "Money doesn't grow on trees," "God is in all things," and "Ask and you shall receive."

What were the mantras you grew up hearing and repeating?

ACTION

It's (journal!) time to create new mantras so that you can create the reality you want to live in.

Make a list of all of the words and phrases you heard as a child. Note: this may not be easy if you blocked aspects of your childhood. If there are blocked aspects of childhood, it means you wanted to forget what happened then. What happened then is happening now in the Quantum. Choose that you won't get up until you remember just one…likely, that one will start a cascade of memories!

Write them down.

Now, write down mantras you say on a regular basis to yourself now. You might be shocked to find that the words and phrases you say to yourself are, to put it politely, less than ideal. This is good work. That unicorn in the corner is proud of you.

You are making real shifts in the Quantum.

Beside each mantra, ask yourself if you want it to stay or go. If it's time to get that energy out of your life, think of new phrases and new words that you can say in your reality NOW to create more of the outcome in the Quantum that you want.

<u>Tip:</u> One of my favorite mantras you are welcome to utilize is, **"Today is the best day of my life."** I say it every single morning, no matter what. I say it to every person I interact with in the morning. In fact, everyone I know now says this mantra: **"Today is the best day of my life."** This mantra creates the foundation of my day every single day, and every day it changes the way my day goes.

Here is one last secret for you: As you evolve in the Quantum, your mantras will evolve as well.

Isn't it lovely?
You're the best.
See you tomorrow.

# *Day* 18
## ALIGNMENT

Your Quantum self fiercely wants to assist your human reality…but it can only come in and assist when your human is in alignment.

Practicing affirmations, staying in a good vibe, and holding the faith can only get you so far. Those efforts connect your emotions and logic. Emotions and logic are only two aspects of alignment.

INTROSPECTION

Alignment is when what you think, say, feel, and do **all match the result.**

How quickly can you identify whether you are in or out of alignment? Are you like a rubber band, able to feel the discomfort of non-alignment and snap right back into your proper form?

Think of when someone complains of being tired. Their words are, "I'm tired"; their feelings are grogginess, weighed down, mental fuzziness; their thought is, "I want to go to bed." But then, what are their actions? And what's the result of their actions?

They might stay up and scroll their phone or watch a movie or read a book. The result is *not going to sleep*. This is not alignment. How do you know? They were tired but did not sleep.

Funnily enough, I am writing this to you now at 4:45 am. I was trying to sleep, tossing and turning. As I laid in bed, I thought, "Why am I trying to sleep when I'm not in alignment? I don't want to sleep. I'm not tired. I'm getting up!"

We're always going to fall out of alignment—that's

part of the contract of being human! It is your awareness and conscious action that snaps you back.

What I think, say, feel, and do *must match the result*. So, I got up and got to work. This is true alignment.

When you are aligned, the Quantum can open to you and guide you to the next best thing. It's easy to live in non-alignment. It's easy to say, "I'm tired," but mindlessly travel down a YouTube rabbit hole. It is not always easy to think, say, feel, and do in alignment, but it is essential for you to get what you want.

It only gets simpler and clearer as you hone your awareness of, *and commitment to*, your alignment.


ACTION

Aligned journal time!

Write ways that you are in total alignment: What you think, say, feel, do, and the result all match. Find examples from your past and your present. Practice, practice!

Did you decide you were going to work out today (thought), say you were going to work out (say), feel like you needed a good sweat (feeling), put on workout clothes (action), and did you do the workout (result)?

This is alignment. Journal it.

Journal the ways you were not in alignment today. Here are some areas to investigate: financial, physical, sexual, emotional, mental, and spiritual. Now write beside them the correct way to be in alignment.

The Quantum is always available if you choose to jump into it—in alignment!!!

Be real. This exercise only works if you are hard on yourself. It might sting a little but do it! Take the time to clearly see how often you are (or are not) living in alignment.

When you are in alignment, true Quantum magic can take place.

Good job, aligned you!
See you tomorrow.

# Day 19

## NONLINEAR HEALING

Your current reality is carrying so much from the past. You are, after all, human for a reason! Your past may have been hard, bad, challenging…and it goes deeper than just your experience.

You carry all that your ancestors went through; specifically, your grandmothers.

Did you know women are born with all the eggs in our bodies for reproduction that we'll ever have… and it gets even more Quantum science magical than this! Within our mother's womb *before we are even conceived,* our mother carries all our eggs, too?! YES. Being human is such a cool trip!

If you are a woman, you repeat cycles and patterns that are trapped in your blood, in the blood of your mother, and her mother. It is why you bleed. (Note: even if you don't physically bleed due to hyster-ectomy or illness, your womb space bleeds in the Quantum. For the purpose of today, believe and know that *you do indeed bleed.*)

Karma isn't bad. It also isn't good. Karma simply is *all that was*. Karma is here for you to work with, and through, in this lifetime.

Karma is an accumulation of experiences; debts and credits. Anything that your ancestors experienced, you are also experiencing in the Quantum. You came as a human to assist in the energetic removal of all that was, and to clean the slate with monthly blood. It is a gift of being a human woman.

Which ancestral wounds would you be ecstatic to clear?
How might it feel to know that you are doing work for all those that came before and all those who will come after you in your bloodline?

You can begin to cleanse these experiences and free yourself and your ancestors of past baggage. Anything is available to you in the Quantum, so today you go into the Quantum to release these experiences from your past. Free up that space and free up the energy held there!

You can create a space right now where you can heal the past and create a new, clean future.

In the Quantum, ALL is available. The Quantum says that you can change the energy in all directions of time, space, reality, and dimensions.

To be clear, you don't want to change what happened because it's bad, you want to change what happened in order to set the tone for what's to come.

Experiences are not bad or good—*they simply are*. They are what had to happen to get you here. And now you get to decide what happens now to get to the next stage.

What you do today doesn't just affect those that come after us, but also affects those that came before us.

There is nothing linear about the Quantum.

*Journal time!*

Turn off your phone and all ways that you can be distracted.

Get into full alignment with your thoughts, words,

<u>actions, and feelings regarding doing this work.</u>
Coming into a space of true alignment for your
WHY creates an opening for transformation.

What thought are you thinking about doing this
work?
What feeling do you have about doing this work?
What words are you saying about doing this work?
What action are you taking in this work?
And the result is that you do the work! Alignment!

<u>Then, set a timer for 11 minutes and let your pen
write down events from the "past" that you do
not want to create more of.</u> A few examples: being raised in poverty, experiencing divorce, abuse,
trauma, etc. These events aren't bad. They happened to get you here...and now you're going to
release them so they are not repeated.

<u>Then write down all the experiences, feelings,
thoughts, and words you want more of.</u> These are
events like completing a degree, starting a business,
being a great mother, and so on.
(These should feel *so good* in your body! Like, "*yes,
more of this please!*")

*When you feel complete, burn the paper.*

Let it go into the Quantum. Brava!
See you tomorrow!

# *Day* 20

## INVITING YOUR SPIRIT
## IN FULLY

When your spirit is turned on, when your spirit clicks into place, your mind and body hum in tune. It all begins to work effortlessly.

This doesn't mean life is effortless. It means that it feels turned-on, lit up, effortlessly full of experiences!

Life is full of effort (love, money, creation), but it feels effortless when aligned.

This is the Quantum.

Your spirit says Do That or Don't Do That. Your spirit tells you how to behave and how to make choices.

Your spirit speaks to you.

You cannot make wrong decisions.

EVER.

Today is the youngest you'll ever be. Today is the first, last, and best day of your life.

You came here to have an *entire* experience. Not an only good experience. Not an only easy experience. Not an only effortless experience.

You came here for the ups and downs, the highs and lows, the goods and bads. You came here for the entire experience.

There are no wrong decisions. Each decision is taking you closer to your spirit. When you learn to listen to your spirit and tune to the Quantum, *your decisions become clearer and clearer*. Sometimes they are easier and sometimes they are harder. Duh! This is being a spirit with a human body!

Nonetheless, the decision is **never** wrong. Isn't that a relief?

Today you invite your spirit back into your body.

Watch the video for today to learn to invite in your spirit.

<u>Journal your process of inviting in your spirit. Write: I AM ME-FOCUSED.</u>

Open your spiritual heart. Swing the doors wide open. Maybe journal this process or meditate with it or sing it or dance it. Open the doors and let your spirit in. Call her in. After she comes in, leave the doors open for your spirit to go fly free.

She is flying around with that jet-packed unicorn in the Quantum multiverse, waiting to come back in to guide you.

Carry your spirit around as a secret. This is part of the feminine mystique. Your spirit is your greatest ally and hidden mystery.

You have all you need inside.

# *Day* 21
## BEGIN AGAIN

See you tomorrow.
Every morning we get to begin again.

Every sentence is the chance to begin again.

Every time you get dressed is another beginning.

Every kiss is a brand-new kiss.

Every shower is a new experience.

Today the practice is to begin again consciously throughout the day. Every single day. All day long!

## INTROSPECTION

Do you walk through your day like a robot or a zombie, doing the same practices and having the same experiences? Do you wake up the same way? Pee the same way? Talk to your dog the same way? Walk the same way? Eat the same way?

Most people do. They live the same life day after day. The worst part is, they think they like it because they've always done it that way. That was the way their mother did it. They haven't asked themselves what they really want.

They don't even know if they want new experiences!! They don't even know exciting new opportunities are waiting in the corner, hoping for an invitation!

It seems so sad to you now, doesn't it? Because you are a Quantum person.

And that old way of living...that is SO NOT QUANTUM!

Quantum Journal time!

Get seriously non-Quantum here and be real with yourself about how you start your day. Write it down. How do you do the first 20 minutes? Then what do you do? And then what?

BORING, am I right?

No more.
How do you want to start your Quantum day? Journal all the ways you can start your day.

Maybe you start one day with a magical entrance! Then the next you proclaim your existence! The next you might karate chop your way into your day! The next maybe you sing your way awake!

Make a bold move to begin your day.

And then all day long, ask yourself how many times you can begin again.

A spirit-led Quantum jet-packed life is about allowing what is possible to surprise you. Possibility is all around you and all you have to do is open to it. This is what beginning again is. Closing and opening your eyes with unending curiosity.

Invite your Quantum self in for fun—because it *is* fun. Re-write your magic formula based on the results.

You get to decide how you begin and how many times you begin...
And isn't that Quantum magic freedom?
I'm proud of you.
You did it.
Now.
Begin again.

# *Day* 22-28

## INTEGRATION

Choice leads to change. Real change.

CAN YOU FEEL IT?!

Think back to Day 1. What were your thoughts? Your beliefs? Your choices?

And now, less than one month later, what are your daily thoughts? Your beliefs? Your choices?

I told you so. And I also told you your life will change as a result. All you ever wanted will begin to come into your life…but it took *first starting with choice*.

CONGRATULATIONS. You made the choice. You followed through. And now...the changes become everyday life.

It's not true that one day you wake up and your life is completely different.

It's true that each day you wake up and you create a completely different life.

Read the following aloud:

*My choices create my reality.*

*I choose to create my reality.*

*I choose to see the potential.*

*I choose to know that just because I can't see something, it still exists.*

*I choose that life is gorgeous, with jet-packed unicorns waiting to surprise me at every turn.*

Now write your own declaration for today.

Each day in the week after "completion" of this workbook is the chance to go deeper into the Quantum by devotion to yourself. Choice. Action. Result.

Come back to this chapter each day in the final week, open your trusty companion journal which has taken you on this wild and brand-new Quantum life. Go through the checklist below to have a CTQ moment (Come to Quantum) with yourself. Journal on one or all of these points to cement the work you've done.

+ How did your day begin today? How was it different than any other day? Did the Quantum open in kind?

+ What does your spirit feel like today? What does she want?

+ How have your thoughts, feelings, words, and actions matched your results today? How have they been out of alignment?

+ Does your heart feel open or shut? What action can take your heart into a Quantum place?

+ What words are you repeating today? What is the mantra in your life?

+ Explain your magic formula and how miracles can play a role in your magic?

+ How has getting to know yourself in silence contributed to your magic?

+ What does your devotion look like now compared to Day 1?

+ What are some of your new norms?

+ How much have you been doing versus receiving?

+ How has healthy fear helped your Quantum self?

+ Compare your energy today to a week ago—what has shifted in you?

+ What do you have? What do you want?

+ How does your body feel? Can you feel your belly fuzz?

+ What physical action has changed your Quantum self the most?

Maybe open the workbook to a wild page and journal on how that day has impacted your life.

*You now live in the Quantum. It is a part of you.*

*You are limitless. Tuned in.*
*Aligned. Expansive. Magnetic.*

I am so proud of you, Quantum friend. You have grown exponentially over these days. Keep growing.

# *Day* 29

## EPILOGUE

I want to tell you something. I hope this inspires you to touch the Quantum daily.

The opening sentence of my first book (which you won't find anywhere—except my Quantum experience—because I trashed it) was: "For as long as I can remember, I have been swept under rugs, stepped over, discarded, pushed aside in lieu of someone or something better…"

Now you see why I trashed it?

I believed I was worthless. I wasn't the prettiest girl. The smartest. The fastest. I wasn't anything special. I was mediocre at best. I had no one to tell me my future could be incredible. Everywhere I looked, I found evidence that I would never become anyone grand based on my past and present.

I was raised in poverty, no father, no grandparents, no trust fund (not even one single vacation as a child), nowhere to go…

*BUT UP.*

You see, I wrote that memoir of my first 27 years…a life I wasn't proud to have.

But a decade later, after implementing what I've taught you within the contents of this workbook, I am now recognized worldwide for leading a team of women to the top .7% of sales, all while growing a human inside my body.

I am easily the happiest person I know in all areas of my life.

All because I have made the choice that I can create whatever I want through my devotion. Choice. Action. Connection to Spirit.

> And the evidence continually
> appears to prove me right.

I pray that through these pages, you have felt my life's work ripple through your own gorgeous existence… and I know your life will become even better *than you have ever imagined*.

And so much more.

**This or something better, unicorn.**
**This and all the things better.**

xx,

*Angela*

www.ingramcontent.com/pod-product-compliance
Lightning Source LLC
LaVergne TN
LVHW052033080426
835513LV00018B/2304